Bei Fragen zur Produktsicherheit wenden Sie sich bitte an:
If you have any questions regarding product safety,
please contact:

Walter de Gruyter GmbH
Genthiner Straße 13
10785 Berlin
productsafety@degruyterbrill.com

Mathematical Research in the last 20 years

Presidential Address
delivered on the 31st January, 1921,
before the Benares Mathematical Society

By

Ganesh Prasad

1923

Berlin und Leipzig
Walter de Gruyter & Co.
vormals G. J. Göschen'sche Verlagshandlung — J. Guttentag, Verlagsbuchhandlung — Georg Reimer — Karl J. Trübner Veit & Comp.

C. G. Röder G. m. b. H., Leipzig. 821223

Contents.

Colleagues and Post-Graduate Students,—

I propose to address you this evening on mathematical research in the last twenty years. But allow me to preface my address with a few remarks on the choice of the subject.

When I stood before the first annual meeting of the Society on this date last year, I discoursed on the functions and needs of a mathematical society and tried to hold before you, so to say, a mirror in which you might see the activities and the methods of work of the mathematical societies of the world. During the last twelve months the progress made by our Society has been steady and its contribution to the advancement of knowledge has been appreciated and praised by some mathematicians of eminence. We are assured of the support of many mathematical and scientific societies in England, Holland, Sweden, Italy, Japan and America which exchange their publications with our journal, and of the sympathy and interest of some great leaders of mathematical thought who have written to me to express their appreciation of the work done by the Society. We are members of a Society which has a bright future and a high promise and I trust our endeavours, if well-directed, will soon secure for the Benares Mathematical Society a distinguished position in the mathematical world. In choosing the subject of my address I have been guided mainly by my desire to give you an idea of what the most important advances in mathematical knowledge were during the twenty years that have just passed away in the present century and so place before you models after which you may shape your own researches.

1. In spite of the vast output of original mathematical papers from year to year, one can see clearly certain important directions in which mathematical knowledge advanced in the period under consideration. Of these directions the most important are those that led to the building up of the theory of integral equations and its applications, the investigation of the foundations of mathematical Physics, the generalization of the notion of convergent series and the development of the principle of relativity.

I. Integral Equations.

2. Although long before 1887 certain integral equations had been studied by Laplace, Abel, Liouville and Joachimsthal, it is in a paper* written by Paul Du Bois-Reymond in October of that year that we find the first use of the term "integral equation", the first mention of a general type of integral equations, *viz.*

$$\int_0^S ds\, f(s)\, \Phi(s, x) = \psi(x)\, f(x) + \chi(x),$$

where S is a constant, Φ, ψ and χ are known functions and f is to be determined, and the first clear expression of the importance of integral equations. The following words† of Du Bois-Reymond are prophetic of the development of the theory of integral equations: "(Since 1852) I have come across integral equations so frequently in the theory of partial differential equations that I feel convinced that the progress of this theory is bound up with the treatment of integral equations about which, however, almost nothing is known. Neither the question, how far the problem (of the solution of integral equations) is a determinate one, nor its classification, as the problem surely appears to be connected with the problem of differential equations, has been, so far as I know, examined as yet". During the years that intervened between 1887 and 1900 when Professor Erik Ivar Fredholm of Stockholm published his first paper on integral equations, a good deal of work was done by Professors Pincherle, Volterra and Levi-Civita.

3. Fredholm's first paper,[1] which appeared in an amplified form[2] in 1903, owes its origin to the following presentation by Poincaré of Neumann's method of solving Dirichlet's problem: —

Let S be a closed plane curve without double points. We have to determine a double layer on S of moment Φ such that its potential W satisfies the equation

$$\frac{1}{2}(W^+ - W^-) - \frac{\lambda}{2}(W^+ + W^-) = F, \tag{A}$$

where W^+ or W^- denotes the limiting value which W takes according

---

* "Bemerkungen über $\triangle z = \frac{\partial^2 z}{\partial x^2} + \frac{\partial^2 z}{\partial y^2} = 0$" (*Journal für die reine und angewandte Mathematik*, Bd. 103).

† Loc. cit., pp. 228-229.

[1] "Sur une nouvelle méthode pour la résolution du problème de Dirichlet" (*Oefversigt af Kongl. vet. akad. Forh. Stockholm, 1900*).

[2] "Sur une classe d'équations fonctionelles" (*Acta Mathematica*, t. 27, 1903).

as the curve is approached from the inside or the outside, and F is the prescribed boundary value for W^+.

Now from the theory of potential, at any point $x = x(s)$, $y = y(s)$ on the curve,

$$W^+ = \pi \Phi(s) + \int_0^l \Phi(t) \frac{\partial}{\partial t}\left\{\tan^{-1} \frac{y(t)-y(s)}{x(t)-x(s)}\right\} dt,$$

$$W^- = -\pi \Phi(s) + \int_0^l \Phi(t) \frac{\partial}{\partial t}\left\{\tan^{-1} \frac{y(t)-y(s)}{x(t)-x(s)}\right\} dt,$$

l being the length of the curve and $\Phi(t)$ being the moment of the double layer. Substituting in (A) we get an integral equation of the type discussed by Fredholm, *viz.*

$$\Phi(s) - \lambda \int_a^b \Phi(t) K(s, t)\, dt = f(s). \tag{B}$$

If we put

$$K(s, t) = \frac{1}{\pi} \frac{\partial}{\partial t}\left\{\tan^{-1} \frac{y(t)-y(s)}{x(t)-x(s)}\right\},$$

$$f(s) = \frac{F(s)}{\pi}, \quad a = 0, \quad b = l, \quad \lambda = -1,$$

the solution of Dirichlet's problem is obtained from the solution of (B).

4. The method of iteration used by Neumann and Volterra gives the solution of (B) in series of powers of λ. As this series is convergent only for certain values of λ, Fredholm discovered a much better method which gives the solution as the ratio of two power series which converge for all values of λ. Fredholm's method consists in treating (B) as the limiting case of a system of simple simultaneous algebraic equations. For, let the interval (a, b) be divided into n equal parts each part being of length $\delta = \dfrac{b-a}{n}$, and let

$$x_p = \Phi(s_p)$$

where $s_p = a + p\delta$, p having the values $0, 1, 2, \ldots n$.

Then (B) gives

$$x_p - \lambda \int_a^b K(s_p, t)\, \Phi(t)\, dt = f(s_p),$$

i. e.,

$$x_p - \lambda \lim_{n=\infty} \left\{\delta \cdot \sum_{q=1}^{q=n} x_q K(s_p, s_q)\right\} = f(s_p).$$

If we solve the $(n+1)$ equations

$$x_p - \delta\lambda \sum_{q=1}^{q=n} x_q K(s_p, s_q) = f(s_p)$$

for the unknown quantities $x_0, x_1, \ldots x_n$, and then proceed to the limit, we get

$$\Phi(s) = f(s) + \lambda \int_a^b K(s, t, \lambda) f(t)\, dt \tag{C}$$

where $K(s, t, \lambda) = \dfrac{D(s, t, \lambda)}{D(\lambda)}$, the numerator and denominator being integral functions in λ.

Fredholm proves that if $D(\lambda)$ is not zero, (C) is a solution, and the only solution, of (B), and that in that case the homogeneous equation

$$\Phi(s) = \lambda \int_a^b K(s, t) \Phi(t)\, dt \tag{D}$$

has no solution other than zero. On the contrary if D (λ) is zero, it is proved by Fredholm that (D) has at least one solution different from zero.

5. With integral equations and specially with the homogeneous equation (D) the name of Professor David Hilbert shall ever remain associated. With a more comprehensive grasp of the possibilities of integral equations than any of his contemporaries, Hilbert undertook in a series of six papers[3] published in the period 1904—1910 to build up in integral equations an instrument which would be more useful and of more general applicability than differential equations. In his first two papers which appeared in 1904, after deducing Fredholm's results by starting from the problem of the orthogonal transformation of n variables and then proceeding carefully to the limit for $n = \infty$, Hilbert lays down a theory of characteristic numbers (Eigenwerte) and characteristic functions (Eigenfunktionen) for the solution of boundary value problems in the theory of differential equations. For example, taking a real self-adjoint system of the second order, *viz.*

$$\frac{d}{ds}\left(k \frac{d\Phi}{ds}\right) + (\lambda - l)\Phi = 0 \tag{E}$$

with boundary conditions in which the parameter λ does not enter, Hilbert showed (*a*) that the system is equivalent to the homogeneous equation (D), the kernel K (s, t) being the Green's function for the equation

$$\frac{d}{ds}\left(k \frac{d\Phi}{ds}\right) - l\Phi = 0,$$

[3] "Grundzüge einer allgemeinen Theorie der linearen Integralgleichungen" (*Nachrichten der Königlichen Gesellschaft der Wissenschaften zu Göttingen*).

(*b*) that the kernel is symmetrical with respect to *s* and *t*, (*c*) that, consequently, there are an infinite number of real roots of $D(\lambda) = 0$, (*d*) that, corresponding to each of these roots, called characteristic numbers of the kernel, there exists at least one characteristic function Φ which is the solution of (D) as well as of (E), (*e*) that all these characterstic functions form an orthogonal system and (*f*) that every function $g(s)$, representable with the help of a continuous function $h(s)$ by the integral

$$g(s) = \int_a^b K(s, t)\, h(t)\, dt,$$

is expansible in an absolutely and uniformly convergent series of characteristic functions, the coefficients being obtained in Fourier's manner. The afore-said results have been generalized in various directions by Hilbert himself in his fifth paper which appeared in 1906, by Professor Erhard Schmidt[4] and many other pupils of Hilbert including Professors Myller, Westfall, Hilb, Haar and Mason[5], and by Professors Kneser, Böcher, A. C. Dixon, Stekloff and Picard[6].

6. In the domain of partial differential equations, Hilbert's work has been equally successful. The result obtained by him in his second paper, that the solution of the equation of elliptic type

$$\frac{\partial^2 \Phi}{\partial x^2} + \frac{\partial^2 \Phi}{\partial y^2} + a\frac{\partial \Phi}{\partial x} + b\frac{\partial \Phi}{\partial y} + c\Phi = f$$

which vanishes on the boundary is also the solution of the integral equation

$$\Phi(x, y) + \frac{1}{2\pi}\iint\left\{\frac{\partial(aG)}{\partial\xi} + \frac{\partial(bG)}{\partial\eta} - cG\right\}\Phi(\xi, \eta)\, d\xi\, d\eta$$

$$= -\frac{1}{2\pi}\iint f(\xi, \eta)\, G(x, y; \xi, \eta)\, d\xi\, d\eta,$$

where G is Green's function relating to

$$\frac{\partial^2 \Phi}{\partial x^2} + \frac{\partial^2 \Phi}{\partial y^2} = 0,$$

---

[4] "Zur Theorie der linearen und nichtlinearen Integralgleichungen" (*Mathematische Annalen*, Bde. 63, 64, 65).

[5] See the Göttingen dissertations of Myller, Westfall and Haar, Hilb's papers in *Crelle's Journal*, Vol. 140 and *Math. Ann.* Vol. 71, Haar's paper in *Math. Ann.*, Vol. 71 and Mason's paper in the *Trans. Am. Soc.*, Vol. 7.

[6] See Kneser's book, "Die Integralgleichungen und ihre Anwendungen in der Mathematischen Physik", Böcher's "Leçons sur les méthodes de Sturm", Dixon's papers in *Phil. Trans.* Vol. 211 (A) and *Proc. L.M.S.*, Series 2, Vol. 7, Stekloff's papers in *Comptes Rendus*, Vols. 150, 151 and his memoir in the memoirs of the Academy of Sciences of Petrograd for 1913, and Picard's papers in *Ann. Ec. norm.* and *C. R.* for 1910 and 1911.

is the starting point of a number of papers by Prof. Leon Lichtenstein[7] who has generalized Hilbert's result in various directions, by Professors Sommerfeld and Carslaw[8], and above all by Professor H. Weyl[9] whose results relating to the asymptotic distribution of the spectrum of a body may be safely characterized as one of the remarkable achievements of the method of integral equations. In his first memoir, says Weyl: "Methods of such generality as the theory of integral equations, when applied to physical problems, accomplish, if I am not mistaken, the discovery of the general features common to a large complex of phenomena. In this sense I have placed before myself the task to prove in this paper the following theorem with the methods of the theory of integral equations: All vibration-phenomena governed by a linear differential equation of the type of the ordinary vibration-equation possess essentially one and the same spectrum in the region of the high vibration-frequencies, independently of the geometrical shape and physical constitution of the bodies in which they are observed". In his second memoir Weyl extends the aforementioned result, obtained for membranes, to elastic vibrations.

7. Hilbert's third paper, which appeared in 1905, is devoted to an application of the theory of integral equations to the solution of the following problem first formulated by Riemann: Determine a regular analytic function of a complex variable for a given region, when certain relations between the real and imaginary parts of its value on the boundary are prescribed. Hilbert's results were generalized by one of his pupils, Professor Haseman[10].

8. In his fourth paper, published in 1906, Hilbert transforms the two fundamental theories of integral equations, viz., Fredholm's theory of (B) and his own theory of characteristic numbers relating to (D) with symmetrical kernel, into problems of a theory of infinite variables. It is with the help of such a theory that Hilbert discusses in his fifth paper integral equations which he calls polar and for which the kernel is of the form

$$p(s)\,H(s,\ t)\,.$$

---

[7] "Zur Theorie der linearen partiellen Differentialgleichungen zweiter Ordnung des elliptischen Typus" (*Math. Ann.*, Bd. 67). Also see his papers in *Acta Math.*, t. 36, 1913, and *Math. Zeitschrift*, Bd. 3, 1919.

[8] See Sommerfeld's paper in *Jahresbericht d. d. Math.-Vereinigung*, Bd. 21, 1913, and Carslaw's paper in *Proc. L.M.S.*, Vol. 13, 1914.

[9] "Das asymptotische Verteilungsgesetz der Eigenwerte linearer partieller Differentialgleichungen" (*Math. Ann.*, Bd. 71, 1912). Also see his paper in *Rendiconti C.M.P.*, t. 39, 1915.

[10] "Anwendung der Theorie der Integralgleichungen auf einige Randwertaufgaben der Funktionentheorie" (*Math. Ann.*, Bd. 66, 1909). Also see his Göttingen dissertation of 1907.

Round the fourth paper of Hilbert have sprung up a number of interesting works by Hellinger, Toeplitz, Frechet, Stäckel, Friedrich Riesz, von Koch, Perron and Lichtenstein[11] dealing with the theory of infinite number of variables, the theory of infinite number of quadratic forms and the theory of linear equations with infinite number of unknowns, and with the applications of these theories to integral equations.

9. Hilbert's sixth paper treats of boundary value problems for a simultaneous system of partial differential equations of the first order and of elliptic type, of a new method, called the method of parametrix, for transforming differential equations into integral equations and *vice versa*, and the application of this method to the most general partial differential equations of the second order and of elliptic type, and concludes with applications of integral equations to Minkowski's theory of volumes and surfaces, a problem in the theory of automorphic functions and lastly a certain two-parameter boundary value problem intimately connected with Klein's oscillation theorem.

10. For integral equations of the type

$$f(s) = \int_a^b K(s, t)\, \Phi(t)\, dt, \tag{F}$$

we owe to Hilbert the result that, when $K(s, t)$ is the Green's function corresponding to certain boundary conditions for a self-adjoint linear differential equation of the second order and $f(s)$ satisfies the same boundary conditions and possesses a continuous second derivate, Φ is given by operating on f with the operator of the given differential equation. It is well-known that (F) is not soluble in general. In fact it has been proved by Professor Bateman in a paper[12] written at Göttingen in 1906 under the influence of Professor Hilbert that in order that (F) may have a solution it is necessary that $f(s)$ be expansible in an absolutely and uniformly convergent series of the characteristic functions corresponding to $K(s, t)$. Among the other investigators of (F) are Professors Lalesco, Lauricella and Picard[13], the last of whom has given the elegant theorem that the necessary and sufficient condition for the existence of a solution of (F) is

---

[11] See Hellinger's paper in *C. J.*, Vol. 136, the papers jointly written by Hellinger and Toeplitz in *C. J.*, Vol. 144 and *Math. Ann.*, Vol. 69, Frechet's Paris Thesis of 1910, Stäckel's paper in *Acta Math.*, t. 37, 1914, Riesz's paper in *Gött. Nachr.* for 1910, von Koch's papers in *Rendiconti C. M. P.*, t. 28 and *Math. Ann.*, Vol. 69, Perron's paper in *Acta Math.*, t. 37 and Lichtenstein's in *Rendiconti C. M. P.*, t. 38.

[12] "The inversion of a definite integral" (*Math. Ann.*, Bd. 63).

[13] See Lalesco's paper in *Journal de Math.* for 1908, Lauricella's papers in *Rendiconti Ac. Lincei* for 1908, 1909 and 1910, and Picard's paper in *Rendiconti C. M. P.*, t. 29.

the convergence of the series $\Sigma\lambda_n^2 a_n^2$ where a_n denotes the Fourier coefficient of $f(s)$ relating to $\psi_n(s)$, ψ_n being the solution of

$$\psi_n(s) = \lambda_n \int_a^b K(s, t)\, \psi_n(t)\, dt.$$

11. Professor Vito Volterra has built up in a series of papers[14] published in 1909—1914 a theory of integro-differential equations, *i. e.*, equations which involve not only the unknown functions under signs of integration but also the unknown functions themselves and their derivates. One of the simplest integro-differential equations is of the type

$$x \frac{\partial V(x, y)}{\partial x} + \lambda V(x, y) + \int_0^y S(\xi, y)\, V(\xi, y)\, d\xi = f(x, y),$$

where S and f are known functions, λ is a constant and V is to be determined. Volterra has applied integro-differential equations to the mechanics of heredity, *e. g.* hysteresis. Among others who have contributed to the subject of integro—differential equations may be mentioned Professors P. Lévy, Andreoli, Crudeli, Evans, Fubini, Horn, Pérés, Schlesinger, Sinigallia and Zeilon[15].

12. Among the important results that have not been referred to in the preceding articles may be mentioned (*a*) those of Professor Andreoli[16] on equations of the type

$$\Phi(s) - \lambda \int_0^{g(s)} K(s, t)\, \Phi(t)\, dt = f(s),$$

where $g(s)$ is a given function of s, (*b*) Professor Whittaker's[17] interesting discoveries that the functions associated with the elliptic cylinder are solutions of

$$\Phi(s) = \lambda \int_{-\pi}^{\pi} e^{ik \sin t \sin s}\, \Phi(t)\, dt,$$

---

[14] See the *Rendiconti Ac. Lincei* for 1909—1914, and the paper "Sur les équations intégro-différentielles et leurs applications" (*Acta Math.*, t. 35, 1912).

[15] See Lévy's Paris Thesis of 1911, Ardreoli's papers in *Rendiconti A. L.* for 1913 and 1914, Crudeli's papers in *Rendiconti A. L,* for 1909 and 1910, Evans' papers in *Trans. Am. Soc.*, Vol 15, Fubini's paper in *Annali di Matematica* for 1913, Horn's paper in *Jahresbericht d. d. M.—V.*, Bd. 23, 1914, Pérés paper in *Rendiconti C. M. P.*, t. 39, Schlesinger's paper in *Jahresbericht d. d. M.—V.*, Bd. 24, Sinigallia's papers in *Rend. 1st. Lomb.* for 1911 and *Rend. A. L.* for 1915, and Zeilon's paper in *Rend. A. L.* for 1915.

[16] *Rendiconti C. M. P.*, t. 37.

[17] *Proc. Congr. Cambridge*, 1912, and *Proc. L. M. S.*, Series 2, Vol. 14.

and that the Lame's function $E_n^m(s)$ is a solution of

$$\Phi(s) = \lambda \int_{-2K}^{2K} \Gamma_n(k \operatorname{sn} t \operatorname{sn} s)\, \Phi(t)\, dt,$$

(*c*) Professor Villat's[18] results relating to integral equations of the form

$$A(s)\,\Phi(s) + \int_a^b \{\Phi(s) - \Phi(t)\}\, K(s, t)\, dt = f(s),$$

A and f being known functions and Φ being unknown, (*d*) the applications to mathematical Physics by Professors Picard, Hadamard, Korn, Boggio, E. E. Levi, Poincaré and Lauricella[19] and (*e*) the results obtained by Professor Hobson[20] relating to (B) in his endeavour to give a wider range to $f(s)$ and $K(s, t)$ than that. which is usually given to them.

II. Foundations of Mathematical Physics.

13. About the beginning of the current century much impetus was given to the exploration of the foundations of mathematical Physics by two announcements, one relating to the subject of the Beneke prize for the year 1901 and the other relating to the Adams Prize for the year 1903. In March, 1898, the philosophical Faculty of the University of Göttingen announced the subject of the Beneke prize for 1901 in the following terms: "For a long time the principle of continuity or, still more specially, the representation by means of unlimitedly differentiable functions, has been regarded as the generally accepted basis for the treatment of natural phenomena. This basis had been introduced by the inventors of Differential Calculus and Integral Calculus as something self-evident; but the more mathematical research has progressed the more it has been shown that at the root of that basis lay a very large number of tacit assumptions which, considering the ever existent inexactitude of our sense perceptions, we are in no way constrained to adopt. Also at the outset the supposition of the molecular constitution of matter comes in conflict with the aforesaid basis. The Faculty wants a memoir of actually scientific interest which exposes in a generally understandable manner the questions under consideration and submits to a searching test the admissibility or ex-

---

[18] *Acta Math.*, t. 40.

[19] See Picard's papers in *Rendiconti C. M. P.* t. 22 and 37, Hadamard's prize memoir of 1908, Korn's paper in *C. R.* for 1907, Boggio's papers in *Rend. A. L.* for 1907 and 1908, Levi's paper in *Atti Ac. Torino* for 1908, Poincaré's paper in *Rendiconti C. M. P.*, t. 29 and his "Sechs Vorträge über ausgewählte Gegenstände aus der reinen Mathematik und mathematischen Physik," and Lauricella's papers in *Rend. A. L.* for 1906, 1907, 1908, 1910, *Nuovo Cimento* for 1907 and *Ann. di Mat.* for 1907.

[20] *Proc. L. M. S.*, Series 2, Vols. 13 and 14.

pediency of the usual representation. The memoir may lean more to the mathematical side or to the philosophical and psychological side; historical studies are desired but not required." In March, 1901, the University of Cambridge annouced as follows the subject of the Adams Prize which was to be awarded in 1903: "The bearing on mathematical Physics of recent progress in the theory of the representation of discontinuous quantity by series, with special consideration of the logical limitations of the processes involved. In the year 1847 Sir George Stokes published his classical memoir on Fourier's theorem and its applications to Physics. Since that time various advances have been made in the representation of continuous quantity by series, but they have had reference mainly to the theory of pure quantity. It is desired that the essayist should consider this subsequent work from the physical point of view. Thus, for instance, it is a question how far recent investigations in this field may have a bearing on the validity of the application of continuous differential analysis to material media which are presumed to be ultimately discrete or molecular in structure."

14. Although both the prizes remained unawarded their announcement led directly or indirectly to a large number of researches. Some of these researches were encouraged by the clear enunciation of the questions at issue by Prof. Felix Klein in the report[21] which he published in 1901 on the subject of the Beneke prize.

15. In point of time, the first of the aforesaid researches is that embodied in Prof. Ganesh Prasad's memoir[22] which was communicated by Prof. Felix Klein to the Royal Society of Sciences of Göttingen in July, 1903, and which was published in September, 1903, among the memoirs of that society with the imprimatur of Prof. Klein. The theory of conduction of heat, as propounded by Fourier, had been handed down by successive generations of writers of text-books from Poisson to Poincaré with scarcely any alteration. It seems that, during the last century, the attention of mathematicians remained diverted from the logical foundations of the theory by the intricacies of analysis which special problems presented. In his memoir Prasad studies these foundations carefully and exhaustively. He builds up three theories, *viz.* (1) a theory which treats the solid as a continuum with the same properties in all its points, (2) a theory which is on a strictly molecular basis as, unlike Maxwell and Boltzmann,

[21] "Gutachten der Göttinger philosophischen Fakultät betreffend die Beneke-Preisaufgabe für 1901" (*Göttinger Nachrichten*, 1901; *Math. Ann.*, Bd. 55).

[22] "Constitution of matter and analytical theories of heat" (*Abhandlungen der Königlichen Gesellschaft der Wissenschaften zu Göttingen*, Mathematisch—Physikalische Klasse, Neue Folge, Bd. II, No. 4).

Prasad does not treat a molecule as a point but as a finite entity and, starting with definite suppositions relating to the size, shape and oscillations of the molecules, gives results which are professedly inexact, the amount of error being capable of evaluation, and (3) a theory on the basis of matter being aether in motion. Prasad points out that Fourier's theory, which is of the same type as his first theory, is seriously defective. The principal features of Fourier's theory as remodelled by Prasad may be outlined as follows:—The equation of the phenomenon is shown to be not a partial differential equation but a partial integral equation, and attention is drawn to the simple principle that it is physical continuity which is necessary and not the continuity of the form of an analytical expression. Thus, taking the problem of the linear conduction of heat in a slab bounded by impermeable faces $x = -\pi$, $x = \pi$, and denoting the temperature by T (x, t), the partial integral equation

$$\int_{-\pi}^{x} \{T(x', t) - T(x', o)\} dx' = \int_{o}^{t} \frac{\partial T(x, t')}{\partial x} dt', \quad (-\pi < x < \pi),$$

takes the place of Fourier's partial differential equation

$$\frac{\partial T}{\partial t} = \frac{\partial^2 T}{\partial x^2}, \quad (-\pi \leq x \leq \pi, \ t \geq o);$$

again, it is pointed out that T (x, o) need not be derivable from the analytical expression, which equals T (x, t) when $t = o$, by putting $t = o$ in that expression. For example, let the initial temperature $f(x)$ be an even function of x and let T $(x, t) =$ V (x, t) when $t > o$, V (x, t) being equal to

$$\frac{1}{2} a_o + \sum_{1}^{\infty} a_n e^{-n^2 t} \cos n x,$$

where the a's are the Fourier's coefficients of $f(x)$. Then, in the remodelled theory, the convergence of the Fourier's series

$$\frac{1}{2} a_o + \sum_{1}^{\infty} a_n \cos n x$$

has no place. On the other hand the continuity of $T(x, t)$, *i. e.*, the equality of $\lim_{t = +o} V(x, t)$ to $f(x)$, follows at once from the continuity of $f(x)$; and the only restriction, imposed on $f(x)$, in order that $V(x, t)$ should be the solution, is the following: For each value of x, within the interval $(-\pi, \pi)$, the upper limit of $\left|\frac{\partial V}{\partial x}\right|$ in the neighbourhood of $t = o$ is finite. Prasad proves the limited scope of Fourier's theory by showing that this theory can have a meaning only when $\frac{d^2 f}{d x^2}$ exists and is finite for every value of x in the interval $(-\pi, \pi)$. For example, in the simple case of

$$f(x) = x = \frac{\pi}{2} - \frac{4}{\pi} \sum_1^\infty \frac{\cos(2n-1)x}{(2n-1)^2}$$

Fourier's theory fails because, for $t = o$ and $x = o$,

$$\frac{\partial T}{\partial t} = \infty,$$

while $\frac{\partial^2 T}{\partial x^2}$ is non-existent.

16. The next branch of mathematical Physics whose logical foundations were explored with success is the theory of potential, gravitational, electrostatic or magnetic. In this theory the central formula is that given by Green, *viz.*

$$V(x, y, z) = \int \varrho \frac{dv}{r} + \frac{1}{4\pi} \int \left\{ \frac{1}{r} \frac{\partial V}{\partial n} - V \frac{\partial}{\partial n}\left(\frac{1}{r}\right) \right\} d S$$

and the other important formulae are

Poisson's equation, $\frac{\partial^2 V}{\partial x^2} + \frac{\partial^2 V}{\partial y^2} + \frac{\partial^2 V}{\partial z^2} = -4\pi \varrho$,

Coulomb-Poisson's equation, $\frac{\partial V}{\partial n_1} + \frac{\partial V}{\partial n_2} = -4\pi \sigma$

and Helmholtz's equation $W^+ - W^- = -4\pi \sigma$.

During the last century successful attempts were made by a large number of mathematicians to investigate the conditions of the validity of these formulae. But if in this matter a stage of finality has been reached it is mainly due to the researches of Professors H. Petrini, G. Prasad and C. W. Oseen.

17. The results of Petrini will be found in two voluminous memoirs[23] which he published one in the *Acta Mathematica* of 1908 and the other in the *Liouville's Journal* of 1909, and in a number of shorter papers[24] which appeared during the period 1912—1916. The most important result of

---

[23] "Les dérivées premières et secondes du potentiel" (*Acta Mathematica*, t. 31, pp. 127—332); "Les dérivées premières et secondes du potentiel logarithmique" (*Liouville's Journal*, sér. 6, t. 5, pp. 127—223).

[24] See for these papers *Arkiv för Matematik, Astronomi och Fysik*, Bd. 8, No. 24; Bd. 9, No. 17, Bd. 12, No. 8.

Petrini, *viz.* that relating to the condition to be satisfied by ϱ in order that Poisson's equation be valid, is this: Let

$$\xi = x + r\cos\theta\,,\ \eta = y + r\sin\theta\cos\Phi\,,\ \zeta = z + r\sin\theta\sin\Phi\,,$$

$\cos\theta = \mu$, $dv = d\xi\, d\eta\, d\zeta$, $0 < h < a$, a being an arbitrarily small quantity.

Then the integral

$$V = \int \frac{\varrho(\xi,\eta,\zeta)\,dv}{r}$$

taken throughout the region Ω bounded by the surface S, possesses in the point (x, y, z) a finite differential coefficient $\frac{\partial^2 V}{\partial x^2}$ if

$$\operatorname*{Lim}_{h=0} \left[\int_0^{2\pi} d\Phi \int_{-1}^{1} (3\mu^2 - 1)\, d\mu \int_h^a \frac{\varrho\, dr}{r} \right]$$

exists and is finite, with similar conditions for the cases of $\frac{\partial^2 V}{\partial y^2}$ and $\frac{\partial^2 V}{\partial z^2}$. If now the additional condition is satisfied that ϱ is continuous then Poisson's equation holds. The statement of Petrini made in the introduction to his first memoir and reproduced by Prof. Wangerin in his review[25] of it, that attempt has been made in the memoir to investigate the necessary and sufficient condition for the existence of the first and second derivatives of V *only on the supposition that ϱ is finite*, as well as Petrini's generalization of Poisson's equation announced to be always true when the second derivatives of V do not exist, was challenged by Prasad in a series of three papers[26] published in the period 1916—1918. By giving simple examples such as

$$\varrho = \cos\left(\log\frac{1}{r}\right),\quad \varrho = \cos\log\frac{1}{r\mu}\,,$$

for none of which all the derivatives $\frac{\partial^2 V}{\partial x^2}, \frac{\partial^2 V}{\partial y^2}, \frac{\partial^2 V}{\partial z^2}$ exist or Petrini's generalization holds, Prasad makes clear the invalidity of Petrini's claim. Prasad's aim has been to supplement the work of Petrini by considering the case when ϱ has a discontinuity of the second kind and in this he has succeeded by applying a method of his own based on Du Bois-Reymond's

---

[25] See *Fortschritte der Mathematik*, Bd. 39, pp. 819—823.

[26] "On the second derivates of the Newtonian potential due to a volume distribution having a discontinuity of the second kind" (*Bulletin of the Calcutta Mathematical Society*, Vol. 9, 1916), "On the failure of Poisson's equation for certain volume distributions" (*Philosophical Magazine*, Vol. 34, 1917), "On the failure of Poisson's equation and of Petrini's generalization" (*Bull. Cal. M. S.*, Vol. 8, 1918).

Infinitary Calculus. In another series of four papers[27] published in the period 1917—1920, Prasad has criticized Petrini's results relating to surface-distributions.

18. Prof. Oseen's contribution to the theory of potential is a remarkable one although it is contained in a single paper[28] published in 1918. Oseen shows how simply the theory can be exposed if one avoids the introduction of the second derivatives of the potential by substituting the equation

$$\int \frac{\partial V}{\partial n} dS = -4\pi \int \varrho \, dv$$

for Poisson's equation, ϱ being supposed to be finite and integrable and V to be continuous together with its first derivatives throughout the region Ω. Oseen deduces the formula of Green by applying the following theorem which he proves: —

If Φ (P) and ψ (P) are two functions of a point P in Ω, each continuous together with its first derivatives, ϱ (P) is a function which is finite and integrable in Ω, and for every closed surface S inside Ω the equation

$$\int \frac{\partial \Phi}{\partial n} dS = -4\pi \int \varrho \, dv$$

holds, the integrals being taken respectively over the surface and the region enclosed by it, then likewise for every closed surface S inside Ω the equation

$$\int (\mathrm{grad}\, \Phi \times \mathrm{grad}\, \psi)\, dv - \int \psi \frac{\partial \Phi}{\partial n} dS = 4\pi \int \psi \varrho \, dv$$

holds.

19. Prof. Oseen has also investigated the logical foundations of Hydrodynamics. His results were first communicated in a paper[29] read before the Royal Swedish Academy of Sciences in 1910 and then in an address[30] delivered before the second Scandinavian Congress of Mathema-

[27] "On the Newtonian potential due to a surface distribution having a discontinuity of the second kind" (*Rendiconti C. M. P.*, t. 42, 1917), "On a peculiarity of the normal component of the attraction due to certain surface distributions" (*Phil. Mag.*, Vol. 36, 1918), "On the normal derivate of the Newtonian potential due to a surface distribution having a discontinuity of the second kind" (*Bull. Cal. M. S.*, Vol. 9, 1919), "On the potential of a double layer whose strength has a discontinuity of the second kind" (*Proceedings of the Benares Mathematical Society*, Vol. 2, 1920).

[28] "Über einen Satz von Green und über die Definitionen von rot und div" (*Rendiconti C. M. P.*, t. 38).

[29] "Über die Bedeutung der Integralgleichungen in der Theorie der Bewegung einer reibenden, unzusammendrückbaren Flüssigkeit" (*Arkiv f. Mat., Astro. och Fys.*, Bd. 6, 1910).

[30] "Om Integralekvationernas betydelse för hydrodynemiken" (*Beretning om den anden Skandinaviske Matematikerkongresi Kjobenhavn*, 1911).

ticians held at Copenhagen in 1911. Oseen says: "The differential equations of Navier admit of being transformed into integral equations in which the differential coefficients $\frac{\partial u}{\partial t}$, $\frac{\partial v}{\partial t}$, $\frac{\partial w}{\partial t}$, $\frac{\partial^2 u}{\partial x^2}$, $\frac{\partial^2 u}{\partial y^2}$, etc...., $\frac{\partial p}{\partial x}$, $\frac{\partial p}{\partial y}$, $\frac{\partial p}{\partial z}$, do not occur. Indeed it can be shown that the integral equations possess systems of solutions for which these differential coefficients do not at all exist. One must then ask whether these solutions of integral equations, which are not at the same time solutions of the differential equations, possess only a mathematical importance or whether some among them have a physical importance. I undertake the task to show that the latter is the case and that therefore not the differential equations of Navier but the integral equations should be considered as expressing adequately the physical hypotheses which lie at the basis of the theory of the motion of a viscous fluid." Oseen starts by translating Newton's hypothesis, regarding the relation between change of momentum and force, into the equation

$$\varrho \iiint (u_{t+\Delta t} - u_t)\, dx\, dy\, dz$$

$$= \varrho \int_t^{t+\Delta t} d\tau \iiint \left\{ X - \left(u \frac{\partial u}{\partial x} + v \frac{\partial u}{\partial y} + w \frac{\partial u}{\partial z} \right) \right\}_{t=\tau} dx\, dy\, dz$$

$$- \int_t^{t+\Delta t} d\tau \int (P_x \cos n x + P_y \cos n y + P_z \cos n z)\, dS$$

with two similar equations for v and w, the volume and surface integrals extending respectively over the volume enclosed by any surface S and S itself, and then directly deduces the integral equations of the motion without the intermediation of Navier's differential equations. This deduction is of interest from two points of view. "First, obviously the theory gains thereby in generality. Secondly, the theory gains much in simplicity. In order to see this, one need only consider how complicated the considerations are which are necessary to decide under what circumstances the potential functions possess differential coefficients of the second order. These questions and others of the same kind arise also in Hydrodynamics if one starts from Navier's differential equations. If one, however, sets up directly the integral equations of Hydrodynamics, all these questions lose their importance for Hydrodynamics."

20. During the years 1912—1914, Prof. Hilbert published a number of epoch-making papers[31] on the logical foundations of the kinetic theory of gases and the elementary theory of radiation. In his first paper, which is the same as the last chapter of his book on Integral Equations, Hilbert criticises Prof. H. A. Lorentz's researches on the kinetic theory of gases and shows that it is a linear integral equation of the second kind with a symmetrical kernel which forms the mathematical basis of that theory, so that, without the careful study of that equation according to the modern methods of the theory of integral equations, a systematic foundation of the theory of gases is impossible. Starting with the impact formula of Maxwell and Boltzmann, Hilbert builds up a systematic theory such that, in order to obtain the proof of the second law of Thermodynamics and Boltzmann's expression for the entropy of the gas, only a logical carrying out of the mathematical operations prescribed by his method is required. The other papers relate to the elementary theory of radiation. Hilbert points out that every one of the previous proofs of Kirchhoff's law about the relation between emission and absorption is invalid, and that no proof is possible without the use of a linear homogeneous integral equation with a symmetrical kernel. For example, if the velocity of light is the same throughout the region in question, this fundamental integral equation, which enables Hilbert to give a rigorous proof, becomes

$$\eta = \frac{\alpha}{4\pi}\iiint \mathrm{K}\eta(x_1, y_1, z_1)\, dx_1\, dy_1\, dz_1\,,$$

where η and α are respectively the coefficients of emission and absorption at (x, y, z), $\mathrm{K} = \frac{e^{-\mathrm{A}}}{r^2}$, A being $\left|\int_{\mathrm{P}}^{\mathrm{P}_1} \alpha\, ds\right|$ and r the distance P P_1 between (x, y, z) and (x_1, y_1, z_1), and the triple integration is carried out throughout the region filled by matter.

III. Generalization of the notion of convergent series.

21. Although infinite series had been introduced in the 17th century and had come into common use in the 18th century, for a clear definition of the notion of convergency one would search in vain the writings of the

---

[31] "Begründung der kinetischen Gastheorie" (*Math. Ann.*, Bd. 72, 1912); "Begründung der elementaren Strahlungstheorie" (*Jahresbericht d. d. Math. Vereinigung*, Bd. 22, 1913);

"Bemerkungen zur Begründung einer elementaren Strahlungstheorie" (*Göttinger Nachrichten* for 1913) and two other papers on the same subject published in *Gött. Nachr.* for 1914.

mathematicians of those centuries from Newton and Leibniz down to Euler and Lagrange. Reckless use of divergent series led many eminent mathematicians to gross errors. For example, Euler held that

$$\sum_{n=0}^{\infty} \frac{1}{x^n} + \sum_{n=1}^{\infty} x^n = 0,$$

and the statement made by him in 1744, that 0 was expansible in the form

$$0 = \frac{1}{2} + \sum_{n=1}^{\infty} \cos n x ,$$

was upheld by Poisson even as late as 1825. The first to define clearly what constitutes a convergent series was Bolzano and the definition given by Bolzano in 1817 was adopted by Cauchy and Abel to both of whom is due the credit of denouncing in clear and emphatic language the use of divergent series. In a classical memoir* rendered famous by the fact that it contains the first rigorous and complete proof of the binomial theorem, Abel says: "Another procedure, which one frequently comes across in Analysis and which only too often leads to contradictions, is the use of divergent series for the calculation of the numerical values of series. A divergent series can never be equal to a definite quantity; it is merely an expression with certain properties relating to the operations to which the series is subjected. The divergent series may at times be of useful service as symbols to enunciate these or those theorems more briefly; but one is never permitted to put such series instead of definite quantities. If one does this, then one can prove whatever one wishes, impossibilities as well as possibilities." Thanks to the influence of the writings of Cauchy and Abel, the use of divergent series, with the possible exception of asymptotic series, came into discredit during the fifty years following the publication of Abel's memoir, although this consummation was not arrived at without strong opposition from some English mathematicians whose view may be summed up in the words of their representative, De Morgan** "The motto which I should adopt against a course which seems to me calculated to stop the progress of discovery would be contained in a word and a symbol — remember $\sqrt{-1}$." Strange though it may appear, it is a fact that the modern and rigorous theory of divergent series directly or indirectly owes its existence to Cauchy's investigation on asymptotic

* "Untersuchungen über die Reihe $1 + \frac{m}{1} x + \frac{m(m-1)}{1.2} x^2 + \ldots$" (*C. J.*, Bd. 1, 1826.)

** "On divergent series, and various points in analysis connected with them" (*Transactions of the Cambridge Philosophical Society*, 1844, Vol. 8, Part II, p. 183).

series and to Abel's investigations on series, specially the famous continuity theorem of Abel, that if $\sum_0^\infty a_n$ is convergent

$$\lim_{x=1} \sum_0^\infty a_n x^n = \sum_0^\infty a_n ,$$

the converse of this theorem, and the problem to find $\lim_{x=1} f(x)$ if it is known that the a' s are such that, for every positive value of x less than 1,

$$f(x) = \sum_0^\infty a_n x^n$$

is convergent.

In building up the theory of divergent series, the fundamental problem that presented itself was to generalize the notion of convergent series, in other words, to make to correspond to a series, which is divergent according to the accepted definition of convergency, a number such that the substitution of this number for the series in all the calculations in which the series can appear gives results which are always exactly true or almost always exactly true. The first to give a generalization of the notion of convergent series was Professor Georg Frobenius who, in a paper† published in 1880, gave a partial solution of the afore-said problem of Abel by proving that, s_n being $a_o + a_1 + \ldots + a_n$, $\lim_{x=1} f(x)$ exists and equals

$$\lim_{n=\infty} \frac{s_o + s_1 + \ldots + s_n}{n+1}$$

if this latter exists, and thus defined the sum of the series $\sum_0^\infty a_n$ as this limit. In the years that intervened between the publication of Frobenius's paper and 1901, the year in which Professor E. Borel published his book[32] on divergent series, the following definitions of the sum $\sum_0^\infty a_n$ were given by Professors Hölder, Cesaro, Borel and Le Roy: —

Hölder's definition: $\sum_0^\infty a_n = \lim_{n=\infty} s_n^{(k)}$, (G)

where k is a positive integer,

$$s_n^{(0)} = s_n , \quad s_n^{(1)} = \frac{s_0 + s_1 + \ldots + s_n}{n+1},$$

$$s_n^{(2)} = \frac{s_o^{(1)} + s_1^{(1)} + \ldots + s_n^{(1)}}{n+1}, \ldots,$$

$$s_n^{(k)} = \frac{s_o^{(k-1)} + s_1^{(k-1)} + \ldots + s_n^{(k-1)}}{n+1},$$

and the series is said to be summable (H k).

† "Über die Leibnizsche Reihe" (*C. J.*, Bd. 89, 1880).

[32] "Leçons sur les séries divergentes", 1901.

Cesaro's definition: $\sum_0^\infty a_n = \operatorname{Lim}_{n=\infty} \frac{S_n^{(k)} \cdot k!}{n^k}$, (H)

where

$$S_n^{(o)} = s_n\,,\; S_n^{(1)} = s_0 + s_1 + \ldots + s_n\,,$$

$$S_n^{(2)} = S_0^{(1)} + S_1^{(1)} + \ldots + S_n^{(1)}, \ldots, S_n^{(k)} = S_0^{(k-1)} + S_1^{(k-1)} + \ldots + S_n^{(k-1)},$$

and the series is said to be summable (C k).

Borel's first definition: $\sum_0^\infty a_n = \operatorname{Lim}_{x=\infty} \left[e^{-x} \left(\sum_0^\infty \frac{s_n x^n}{n!} \right) \right]$, (K)

and the series is said to be summable (B 1).

Borel's second definition: $\sum_0^\infty a_n = \int_0^\infty e^{-x} A_k(x)\, dx$, (L)

where

$$A_k(x) = (a_0 + a_1 + \ldots + a_{k-1}) + (a_k + a_{k+1} + \ldots + a_{2k-1})\, x$$
$$+ (a_{2k} + a_{2k+1} + \ldots + a_{3k-1}) \frac{x^2}{2!} + \ldots$$

and the series is said to be summable (B2).

Le Roy's definition: $\sum_0^\infty a_n = \operatorname{Lim}_{t=1} \sum_0^\infty \frac{\Gamma(nt+1)}{\Gamma(n+1)} a_n$, (M)

where $0 < t < 1$.

22. The researches that followed the publication of Borel's book may be conveniently divided into four groups according as they treat of (1) new definitions of the sum or the comparison of various definitions, (2) applications to power series, (3) applications to Dirichlet's series or (4) applications to Fourier's series or other series of orthogonal functions.

(1) Among the contributions of this group the following definition of Dr. Marcel Riesz[33] occupies a prominent place on account of its applications to the theory of Dirichlet's series:—

Riesz's definition: $\sum_0^\infty a_n = \operatorname{Lim}_{\nu=\infty} \sum_{n=1}^{\nu} \left(1 - \frac{\lambda_n}{\lambda_\nu}\right)^r a_n$, (N)

in which λ_n is any real function of n which tends to infinity with n. The series is said to be summable (R $r\,\lambda$), *i. e.*, summable according to Riesz's definition by means of type λ and order r. But the most important contribution of this group is the formulation by Professor G. H. Hardy in

[33] See his papers in the issues of the *Comptes Rendus* of June, July and November, 1909.

a paper[34] written in 1902 of the following general principle of summability: "When a number of limit operations, performed in a definite order on a function of several variables, lead to a definite result, but do not do so when performed in another definite order, we are to agree that the expression which is the result of the formal carrying out of the second sequence of operations Means the result of the first sequence." During the period 1902—1911, a large number of papers appeared in which one or more of the definitions (G), (H), (K), (L), (M), (N) were studied with a view to determine their respective ranges of applicability, their equivalence, and suitable consistency theorems. For example, thanks to the investigations of Professors Knopp, Schnee, Ford and Schur[35] we have several proofs of the complete equivalence of (G) and (H), the simplest of these proofs being Schur's. Again the researches of Hardy and Littlewood have given us the remarkable fact that a particular definition may be inapplicable to a series $\sum_0^\infty a_n$ not only when the series is too divergent but also when it is too nearly convergent. For example, (G) and (H) are both inapplicable unless $n^{-k} a_n$ tends to 0, and, if $|n a_n| < K$, a finite quantity, the definitions are inapplicable unless the series is convergent; thus $\sum_0^\infty n^{-1-i\alpha}$ is not summable by (G) or (H) for any value of α. These results and similar results for (N) will be found in two papers, one published by Hardy[36] in 1910 and the other by Littlewood[37] in 1911. In a paper[38] published by Hardy and Littlewood, a number of interesting theorems have been proved for (K), including the following: If $\sqrt{n}\, a_n$ tends to 0, $\sum_0^\infty a_n$ is not summable unless it is convergent. The importance of Hardy's general principle of summability is evidenced by the fact that it has been successfully applied by him and Dr. Chapman[39] for discussing some very general classes of

---

[34] "On differentiation and integration of divergent series" (*Transactions of the Cambridge Philosophical Society*, Vol. 19).

[35] See Knopp's Berlin dissertation of 1907, Schnee's paper in *Math. Ann.*, Vol. 67, Ford's paper in *American Journal of Mathematics*, Vol. 32, and Schur's paper in *Math. Ann.*, Vol. 74.

[36] "Theorems relating to the summability and convergence of slowly oscillating series" (*Proc. L. M. S.*, Ser. 2, Vol. 8).

[37] "The converse of Abel's theorem on power series" (*Proc. L. M. S.*, Ser. 2, Vol. 9).

[38] "The relations between Borel's and Cesáro's methods of summation" (*Proc. L. M. S.*, Ser. 2, Vol. 11). In this connection see Hardy and Littlewood's paper, "Theorems concerning the summability of series by Borel's exponential method" (*Rendiconti C. M. P.*, t. 41, 1916).

[39] "A general view of the theory of summable series" (*Quarterly Journal of Pure and Applied Mathematics*, Vol. 42, 1911). In this connection see Dr. G. Sannia's paper, "Nuovo metodo di sommazione delle serie: estensione del metodo di Borel" (*Rendiconti C. M. P.*, t. 42, 1917).

definitions and thus obtaining a clearer insight into the nature of the definitions (G), (H), (K), (L), (M), (N), coordinating them as particular cases of the general principle, and, thereby, finding in a simple and natural manner interesting generalizations of almost all the theorems referred to above.

(2) Because of the obvious connection between the two groups, many of the papers of the first group are common to this group. But a number of papers have still to be mentioned. In two papers[40] published in 1907, Hardy studies the cases of series $\sum_0^\infty a_n$ which are summable according to one definition or another, but such that the associated function $f(x)$, instead of tending to the sum of $\sum_0^\infty a_n$ as x approaches 1, oscillates between finite limits of indetermination. For example, Hardy proves that the behaviour of $\sum_0^\infty (-1)^n x^{\psi(n)}$ for $x = 1$, $\psi(n)$ being an increasing function of n of a regular and simple type such as

$$n, n^2, n^3, 2^n, 3^n, \ldots n!, (n!)^2, \ldots,$$

is this: — If the increase of $\psi(n)$ is not too rapid, the series tends to the limit $^1/_2$; when the increase of $\psi(n)$ passes a certain limit, the series oscillates and the swing of the oscillations is greater the greater the increase of $\psi(n)$; when the increase of $\psi(n)$ is greater than or equal to that of $n!$ the swing covers the whole interval (0, 1). Another set of papers deal with the converse of Abel's theorem and with Tauberian theorems, *i. e.*, theorems of the nature of the following theorem given by Tauber in 1897: The existence of $\lim_{x=1} \sum_0^\infty a_n x^n$ and the condition $\lim_{n=\infty} n\, a_n = 0$ together involve the convergence of $\sum_0^\infty a_n$. In a paper [37] by Littlewood, already referred to above, we find the result, that Tauber's theorem holds even if, for

$$\lim_{n=\infty} n\, a_n = 0,$$

the less restrictive condition

$$n\,|a_n| < K$$

is substituted. The paper[41] published by Hardy and Littlewood in 1914 deals with Tauberian theorems for the case in which a_n is positive. Such

---

[40] "On certain oscillating series" (*Quarterly J. P. A. M.*, Vol. 38); "Some theorems concerning infinite series" (*Math. Ann.*, Vol. 64).

[41] "Tauberian theorems concerning power series and Dirichlet's series whose coefficients are positive" (*Proc. L. M. S.*, Ser. 2, Vol. 13).

cases were studied in an important paper[42] published in 1904 by Professor Pringsheim who proved that, if, as n tends to infinity

$$s_n \sim A n^{\alpha} L(n)$$

where A differs from 0, the indices $\alpha, \alpha_1, \alpha_2, \ldots$ being real and such that $n^{\alpha} L(n)$ tends to a positive limit or to infinity, $L(u)$ standing for $(\log u)^{\alpha_1} (\log\log u)^{\alpha_2} \ldots$, then

$$f(x) \sim A \frac{\Gamma(\alpha+1)}{(1-x)^{\alpha+1}} L\left(\frac{1}{1-x}\right),$$

as x tends to 1. Hardy and Littlewood's paper establishes the converse of the above theorem. For example, it is proved that if

$\sum_0^{\infty} a_n x^n \sim \frac{A}{(1-x)^{\alpha}}$, α being positive, then

$$s_n \sim \frac{A n^{\alpha}}{\Gamma(1+\alpha)}.$$

(**3**) The first to make a successful attempt at a systematic study of divergent Dirichlet's series with the help of one of the definitions of summability was Professor Harald Bohr who published his first paper[43] on the subject in January, 1909, another[44] in the same year and his elaborate dissertation[45] and a paper[46] in 1910. The series studied by Bohr is the ordinary one, *viz.* $\sum_1^{\infty} \frac{a_n}{n^s}$ and the definition used is (H). Bohr proves that, just as the region of ordinary convergence is a half-plane, so for every value of k there is a line of summability (C k) for the series, *viz.*

$$R(s) = \sigma_k = \overline{\operatorname*{Lim}_{n=\infty}} \left\{ \frac{(S_n^k \, k! \, n^{-k})}{\log n} \right\}.$$

The series is summable (C k) if $R(s) > \sigma_k$, and not so summable if $R(s) < \sigma_k$. For example, in the case of the series $\sum_1^{\infty} \frac{(-1)^{n-1}}{n^s}$, $\sigma_k = -k$, and therefore the series is summable by one or other of Cesaro's means all over the plane. Another result of Bohr's is the following: If λ_k denote the abscissa of summability (C k) then

$$1 \geq \lambda_{k-1} - \lambda_k, \text{ and } \lambda_{k-1} - \lambda_k \geq \lambda_k - \lambda_{k+1},$$

[42] "Über den Divergenz-Charakter gewisser Potenzreihen an der Konvergenzgrenze" (*Acta Math.*, Vol. 28).

[43] "Sur la série de Dirichlet" (*Comptes Rendus*).

[44] "Über die Summabilität Dirichletscher Reihen" (*Gött. Nachr.*).

[45] "Bidrag til de Dirichlet'ske Raekkers Theorie" (Copenhagen).

[46] "Über die Summabilitätsgrenzgerade der Dirichletschen Reihen" (*Sitzungsberichte der k. Akademie der Wissenschaften in Wien*).

so that going from the right to the left we find that the summability-strips do not increase in breadth. Many other results have been given by Bohr, some of which have been generalized by Schnee in a paper[47] written in 1911. Cesaro's means have also been applied by Hardy and Littlewood for the study of the ordinary Dirichlet's series in a paper[48] published in 1912. The afore-said theorem of Bohr about the summability-strips is found there as a corollary, and many other results are given, some of them being Tauberian in character. Tauberian theorems for positive coefficients will be found in the paper[41] by Hardy and Littlewood already referred to in (2). In a paper[49] published in 1910, Hardy obtained a number of results by applying Borel's method of summation to the ordinary Dirichlet's series. Among these results are (*a*) if $\sum_1^\infty a_n$ is summable, so is $\sum_1^\infty \frac{a_n}{n^s}$, where $R(s) > 0$; (*b*) if $\sum_1^\infty a_n$ is absolutely summable, so is $\sum_1^\infty \frac{a_n}{n^s}$, where $R(s) > 0$; (*c*) if $\sum_1^\infty a^n x^n$ is convergent for $|x| < 1$, and the function represented by the series is regular for $x = 1$, then $\sum_1^\infty \frac{a_n}{n^s}$ is summable for all values of s and represents an integral function of s. The definition (N) of Riesz was framed by him as being specially suitable for application to the case of the general Dirichlet's series $\sum_1^\infty a_n e^{-\lambda_n s}$; and his results will be found in the papers[33] published by him in 1909 and in the tract[50] brought out by him and Hardy in 1915.

(4) The first contributions of this group were the two papers[51] which Professor Leopold Féjer published in December, 1900 and April, 1902, respectively and in which he proved his famous result, that the continuity of a function suffices for the summability (C 1) of the Fourier's series

$$\frac{1}{2\pi}\int_0^{2\pi} f(\alpha)\,d\alpha + \sum_{n=1}^{\infty}\left[\frac{1}{\pi}\int_0^{2\pi} f(\alpha)\cos n(\alpha - x)\,d\alpha\right]$$

---

[47] "Über den Zusammenhang zwischen den Summabilitätseigenschaften Dirichletscher Reihen und ihrem funktionentheoretischen Charakter" (*Acta Math.*, Vol. 35).

[48] "Contributions to the arithmetic theory of series" (*Proc. L. M. S.*, Ser. 2, Vol. 11).

[49] "The application to Dirichlet's series of Borel's exponential method of summation" (*Proc. L. M. S.*, Ser. 2, Vol. 8).

[50] "The general theory of Dirichlet's series" (Cambridge Mathematical Tracts, No. 18, 1915).

[51] "Sur les fonctions bornées et intégrables" (*Comptes Rendus*).

corresponding to it. These papers appeared in an amplified form in a Hungarian journal[52] in 1902 and then in the *Mathematische Annalen*[53] in 1903. Hardy[36] proves that, in virtue of his theorem, that a series for which $|n a_n| < K$ is convergent if it is summable (C 1), the convergence of the Fourier's series corresponding to a continuous function satisfying Dirichlet's conditions follows as an immediate corollary from Féjer's result. In the same way Littlewood's extension of Tauber's theorem referred to in (2) enables Littlewood to exhibit[37] Fourier's theorem as a corollary of the result known under the name of "Poisson's integral", thereby completing Poisson's attempted proof of Fourier's theorem. In a paper published in the *Proc. L. M. S.* in 1913, Hardy gives the following remarkable generalization of Féjer's result: the Fourier's series for any summable function is summable (Ck), for any positive value of k, almost everywhere.

In a paper[54] written in September 1908, Féjer extended his method to the Laplace's series

$$\sum_{n=0}^{\infty} \frac{2n+1}{4} \iint f(\theta', \Phi')\, P_n(\cos\gamma) \sin\theta'\, d\theta'\, d\Phi'$$

corresponding to a function $f(\theta, \Phi)$ and proved that, if the function is continuous at every point of the unit sphere, the series is summable (H 2). The result of Féjer has been generalized in certain important directions by Chapman[55], Dr. Gronwall[56] and Haar[57]. Thus Chapman considers summability ($C\,k$) for $k < 2$ and investigates the suitable additional conditions to be imposed on $f(\theta, \Phi)$; Gronwall improves on Chapman's results in a certain sense by proving, among other things, that the series is summable (C 1); Haar shows that Féjer's method is applicable to series of orthogonal functions in general.

IV. Development of the principle of relativity.

23. The principle of relativity which underlies Newtonian mechanics may be enunciated as follows: —

If K is a system of coordinates relative to which Newton's laws of motion hold, then they hold equally good relative to another system of co-

---

[52] *Mathematikai és Physikai Lapok.*

[53] "Untersuchungen über Fouriersche Reihen" (*Math. Ann.*, Vol. 58).

[54] "Über die Laplacesche Reihe" (*Math. Ann.*, Vol. 67).

[55] "On the general theory of summability, with applications to Fourier's and other series" (*Quarterly J. P. A. M.*, Vol. 43, 1912).

[56] "Über die Laplacesche Reihe" (*Math. Ann.*, Vol. 74, 1913). "Über die Summierbarkeit der Reihen von Laplace und Legendre" (*Math. Ann.*, Vol. 75, 1914).

[57] "Zur Theorie der orthogonalen Funktionensysteme" (*Math. Ann.*, Vol. 69, 1910). In this connection see Dr. C. N. Moore's paper, "The summability of the developments in Bessel functions with applications" (*Transactions of the American Mathematical Society*, Vol. 10, 1909).

ordinates K′ provided that K′ has a uniform motion of translation with respect to K. (P)

This principle, which, in other words, may be expressed as the principle of the covariance of the equations of Newtonian mechanics with reference to the transformation

$$x' = x - vt, y' = y, z' = z, t' = t, \tag{Q}$$

v being the velocity of K′, is dependent on the conceptions of absolute space and absolute time which Newton introduced and which were severely criticized by eminent mathematicians long before the modern theory of relativity was propounded by Prof. Einstein. For example, writing in 1870 Duhamel* says: "The absolute motion generally admitted up to now is a *pure chimera* based on another chimera, that of an eternal and absolute space. We have also to combat a conception, as chimerical as that of space, *viz.* that which makes time an entity, real, necessary, and independent of all creation." Again, Mach** says: — "Newton referred all of mechanics to absolute space! Newton is indeed a gigantic personality; little worship of authority is needed to succumb to his influence. Yet even his achievements are not exempt from criticism. It appears to be pretty much one and the same thing whether we refer the laws of motion to *absolute* space, or enunciate them in a perfectly *abstract* form; that is to say, without specific mention of any system of reference. But owing to the fact that the first course, wherever there was any real issue at stake, was nearly always interpreted as having the same meaning as the latter, Newton's error was fraught with much less danger than it would otherwise have been, and has for that reason maintained itself so long. It is psychologically and historically intelligible that in an age deficient in epistemological critique empirical laws should at times have been elaborated to a point where they had no meaning. It cannot therefore be deemed advisable to make metaphysical problems out of the errors and oversights of our scientific forefathers, but it is rather our duty to correct them, be they small people or great." Notwithstanding the dissatisfaction with the Newtonian conceptions of space and time, the validity of (P) was unquestioned during the whole of the last century, and it was firmly believed that by purely internal mechanical experiment and observation, *i. e.*, not looking outside to external systems, one could never be aware of the earth's annual motion which, for a sufficiently short period, say a few days, may be taken to be very approximately one of uniform translation. But physicists hoped to detect this motion by means of purely ter-

---

* "Des méthodes dans les sciences de raisonnement", Vol. 4, p. 454.

** "Die Mechanik in ihrer Entwicklung historisch kritisch dargestellt", 4th edition, 1901, pp. 570--571 of Dr. T. J. Mc Cormack's translation.

restrial optical experiments. As Maxwell's researches had shown all optical phenomena to be of electromagnetic origin, it was necessary to investigate carefully the equations of electrodynamics for moving bodies. Guided by his belief in the validity of (P) for electrodynamical phenomena, in 1890 Hertz obtained what he believed to be a better set of equations for moving bodies than those given by Maxwell. But both the sets of equations were found to be in full disagreement with experiments. For example, the dragging coefficient of Fresnel turned out to be $\frac{1}{2}$ or 1, instead of $1-\frac{1}{n^2}$, according as one took Maxwell or Hertz to be right. The first successful attempts to smooth out this discordance from experiments were made by Professor H. A. Lorentz. In a paper† which was published in 1892, the electrodynamical theory of Lorentz, generally called the Electron Theory, was proposed, and was given in a simplified and extended form in his famous paper†† of 1895. Postulating a stationary, homogeneous and isotropic aether, Lorentz gave the fundamental equations

$$\left.\begin{aligned} &\frac{\partial \mathbf{E}}{\partial t} + \varrho \mathbf{p} = c \operatorname{curl} \mathbf{M}, \text{ where } \varrho = \operatorname{div} \mathbf{E}; \\ &\frac{\partial \mathbf{M}}{\partial t} = - c \operatorname{curl} \mathbf{E}; \operatorname{div} \mathbf{M} = 0; \end{aligned}\right\} \qquad \text{(R)}$$

which hold exactly or microscopically throughout the whole space, and the macroscopic equations

$$\left.\begin{aligned} &\frac{\partial \mathscr{E}}{\partial t} = c \operatorname{curl} \mathbf{M}'; \operatorname{div} \mathscr{E} = 0; \\ &\frac{\partial \mathbf{M}}{\partial t} = - c \operatorname{curl} \mathbf{E}', \operatorname{div} \mathbf{M} = 0; \\ &\mathbf{M}' = \mathbf{M} - \frac{1}{c} \mathrm{V} \mathbf{v} \mathbf{E}'; \\ &\mathrm{K} \mathbf{E}' = \mathscr{E} + \frac{1}{c} \mathrm{V} \mathbf{v} \mathbf{M}, \end{aligned}\right\} \qquad \text{(S)}$$

obtained by a process of averaging, and neglecting terms of the same order as $\frac{v^2}{c^2}$ and of lower orders, v being the velocity of the body and c that of light. Lorentz's equations were found to be in agreement with experiments so far as first order influence was concerned. But they failed

† "La théorie électromagnétique de Maxwell et son application aux corps mouvants" (*Archives Néerlandaises*, Vol. 25).

†† "Versuch einer Theorie der elektrischen und optischen Erscheinungen in bewegten Körpern" (E. J. Brill, Leyden).

to explain the negative result of the famous experiment of Professors Michelson and Morley because this was a second order experiment. In order to explain this result on the basis of (S), Lorentz was led to the hypothesis that a body contracts by $\frac{1}{2}\frac{v^2}{c^2}$ per unit length in the direction of its motion on account of its motion through the aether with velocity v. But this hypothesis was found to be insufficient for explaining the negative results of the experiments of the late Lord Rayleigh and Professor Brace, performed in 1902 and 1904 respectively to detect double refraction expected by them to be due to the contraction. To explain these results, Lorentz undertook a discussion[58] of the electromagnetic phenomena in a uniformly moving body without neglecting terms of any order and was led in 1904 to the famous transformation known as the Lorentz transformation, *viz.*

$$x' = \gamma(x - vt),\ y' = y,\ z' = z,\ t' = \gamma\left(t - \frac{v}{c^2}x\right), \tag{T}$$

$$\gamma \text{ standing for } \frac{1}{\sqrt{1 - \frac{v^2}{c^2}}},$$

and the result that for free aether (R) is covariant with respect to this transformation. The modified theory of Lorentz, represented by (T), satisfied the requirements of self-consistency and accounted for the negative results of all purely terrestrial experiments intended to show the earth's annual motion.

24. The principle of relativity, which Einstein propounded in his paper[59] published in 1905, may be enunciated as follows:

Every general law of physical phenomena must be so constituted that it is transformed into a law of exactly the same form, when, instead of the variables x, y, z, t of the original (Galileian) co-ordinate system K, we introduce new variables x', y', z', t', of a co-ordinate system K′ connected with K by (T); or, in orther words, every general law of physical phenomena is covariant with respect to the Lorentz transformation.

The aim of Einstein in writing his paper was to introduce such fundamental ideas on space and time as would establish a perfect reciprocity or equivalence of two co-ordinate systems, K and K′, each having a motion of uniform translation in relation to the other. Einstein begins with a critical analysis of the concept of simultaneity and, *assuming* that every

---

[58] "Electromagnetic phenomena in a system moving with any velocity smaller than that of light" (*Proceedings of the Royal Academy of Amsterdam*, Vol. 6, 1904).

[59] "Zur Elektrodynamik bewegter Medien" (*Annalen der Physik*, Series 4, Vol. 17, pp. 891—921).

light disturbance is propagated, in vacuo, relatively to K with a determinate velocity c, no matter whether the disturbance is emitted from a source stationary in or moving with respect to K, he *deduces* that x', y', z', t', the variables corresponding to an event when referred to K', must be connected by (T), with x, y, z, t, the variables corresponding to the same event when referred to K, the velocity of uniform translation being v and its direction that of the axes of x and x'. Some general results of Einstein's theory, called the special theory of relativity, may be enumerated as follow: — (a) In contradistinction with what Newtonian mechanics teaches, neither the time-interval (time) between two events nor the space-interval (distance) between two points of a rigid body is independent of the condition of uniform motion of the co-ordinate system, the former increasing, and the latter decreasing, on account of the motion; (b) the Newtonian vector formula $\mathbf{v}_1 + \mathbf{v}_2$ for the composition of two velocities is replaced by Einstein's famous formula

$$\left[\frac{(\mathbf{v}_1 + \mathbf{v}_2)^2 - \frac{1}{c^2}(\mathrm{V}\,\mathbf{v}_1\,\mathbf{v}_2)^2}{\left\{1 + \frac{1}{c^2}(\mathbf{v}_1\,\mathbf{v}_2)\right\}^2}\right]^{\frac{1}{2}};$$

(c) a point moving uniformly with respect to K moves uniformly with respect to K', but the Newtonian equations of motion of an accelerated particle are not strictly covariant with respect to (T); (d) the kinetic energy of a material particle of mass m is no longer given by the well-known expression $\frac{m\mathrm{V}^2}{2}$ but by the expression

$$\frac{mc^2}{\sqrt{1 - \frac{\mathrm{V}^2}{c^2}}}$$

thereby leading to the important result that the inertial mass of a body is not a constant but varies according to the change in the energy of the body. The theory given by Einstein in his first paper is developed more fully in a paper[60] published by him in 1907 which contains an exposition of the dynamics of a particle according to the theory of relativity. Another important paper[61] of Einstein deals with the inertia of energy; he shows that if a body takes up an amount of energy E_0 then its inertial mass

[60] "Über das Relativitätsprinzip und die aus demselben gezogenen Folgerungen" (*Jahrbuch der Radioaktivität und Elektronik*, Vol. 4, pp. 411—462).

[61] "Über die vom Relativitätsprinzip geforderte Trägheit der Energie" (*Ann. d. Phys.*, Series 4, Vol. 23, 1907).

is increased by $\frac{E_0}{c^2}$; thus if we look upon the inertial mass of a system of bodies as a measure of the energy of the system, the law of the conservation of mass becomes identical with the law of the conservation of energy.

Among those whose researches contributed towards the development of Einstein's theory, Minkowski occupied a prominent place. His chief merit is the discovery[62] published by him in 1907 that "the four dimensional space-time continuum of the theory of relativity, in its most essential formal properties, shows a pronounced relationship to the three-dimensional continuum of Euclidean geometrical space." For showing this, Minkowski put $x_1 = x$, $x_2 = y$, $x_3 = z$, $x_4 = \sqrt{-1}\, ct$, with similar equations for x', y', z', t', so that $x_1^2 + x_2^2 + x_3^2 + x_4^2 = x'^2_1 + x'^2_2 + x'^2_3 + x'^2_4$, and (T) corresponds to a "rotation" of the co-ordinate system in the four-dimensional continuum, described by the co-ordinates x_1, x_2, x_3, x_4, and called by him the world, the rotation being through an imaginary angle $\tan^{-1} \frac{v\sqrt{-1}}{c}$ and in the plane of x_1 and x_4. By this representation and another given by Minkowski in his famous address[63] on space and time, the theory of Einstein gained much in clearness. Interesting graphic representations for the composition of velocities have been given by Sommerfeld[64], Prof. V. Varićak[65] and Dr. A. A. Robb[66]. Various branches of the mechanics of continua have been treated by many writers in the light of Einstein's theory, *e. g.* Hydrodynamics by Dr. Lamla[67], the theory of Elasticity by Prof. Herglotz[68] and Thermodynamics by Prof. Planck[69].

25. Realizing that there is nothing in Newtonian mechanics or in the special theory of relativity to justify the privileging of co-ordinate systems K and K′ for the formulation of the general laws of physical pheno-

---

[62] See his paper "Die Grundgleichungen für die elektromagnetischen Vorgänge in bewegten Körpern" (*Gött. Nachr.*).

[63] "Raum und Zeit" (*Physikalische Zeitschrift*, Vol. 10, 1909).

[64] See his paper in the *Verhandlungen der deutschen Physikalischen Gesellschaft*, Vol. 11, 1909, p. 577; in this connection also see his paper "Zur Relativitätstheorie" in *Ann. d. Phys.* for 1910.

[65] See his papers in the *Phys. Zeit.*, Vol. 11, 1910; in this connection also see his memoir in *Jahresbericht d. d. M.—V.*, Vol. 21, 1912.

[66] "Optical Geometry of Motion" (W. Heffer and Sons, Cambridge, 1911).

[67] "Über die Hydrodynamik des Relativitätsprinzips" (*Ann. d. Phys.*, Vol. 37, 1912).

[68] See his paper in *Ann. d. Phys.*, Series 4, Vol. 36, 1911.

[69] "Zur Dynamik bewegter Systeme" (*Ann. d. Phys.*, S. 4, Vol. 26, 1908, pp. 1—34).

mena, Einstein built up, in a series of papers[70], published during the period 1911—1915, and summarized in his famous memoir[71] of 1916, a general theory of relativity based on the following two principles:

General Principle of relativity: All (Gaussian) coordinate systems are essentially equivalent for the formulation of the general laws of physical phenomena. (V)

Principle of equivalence. A possible alteration, which an observer perceives in the course of an event as due to the action of a gravitational field, would be perceived by him to be exactly the same, if the gravitational field were non-existent and the co-ordinate system used by the observer at his place of observation were to move with a suitable acceleration. (W)

The outstanding result of the general theory is Einstein's law of gravitation which is preferable to Newton's law not only because this law had become ambiguous on account of the new conceptions of distance and inertial mass introduced by the special theory of relativity but also because it has shown itself to be powerless against a recent astronomical observation[72] which confirms Einstein's law. Taking any four Gaussian coordinates x_1, x_2, x_3, x_4 instead of x, y, z, t, and the line-element ds given by

$$ds^2 = \sum_{\nu=1}^{4} \sum_{\mu=1}^{4} g_{\mu\nu}\, dx_\mu\, dx_\nu\,,$$

$g_{\mu\nu}$ being equal to $g_{\nu\mu}$, Einstein's law is embodied in the ten partial differential equations of the second order which the ten g's must satisfy; these equations are typified by

$$G_{\mu\nu} = 0 \text{ or } K_{\mu\nu} \qquad \text{(X)}$$

according as the region considered is empty space or not, where the contracted Riemann-Christoffel tensor

---

[70] The principle of equivalence was given by Einstein in 1911 in his paper in *Ann. d. Phys.*, Vol. 35, pp. 898 et seq. With Prof. Grossmann, Einstein published "Entwurf einer verallgemeinerten Relativitätstheorie und einer Theorie der Gravitation" (*Zeitschrift für Mathematik und Physik*, Vol. 62, 1913) and "Kovarianzeigenschaften der Feldgleichungen der auf die verallgemeinerte Relativitätstheorie gegründeten Gravitationstheorie" (*Z. f. Math. u. Phys.*, Vol. 63, 1914). Einstein communicated two papers, entitled "Zur allgemeinen Relativitätstheorie", to the Berlin Academy of Sciences on the 11th and 25th Nov. 1915. (See pp. 799—801 and 844—847 of the *Sitzungsberichte* of that Academy.)

[71] "Grundlagen der allgemeinen Relativitätstheorie" (*Ann. d. Phys.*, S. 4, Vol. 49, pp. 769—822).

[72] See Dr. Harold Jeffrey's paper "On the crucial test of Einstein's theory of Gravitation" (*Monthly Notices of the Royal Astronomical Society*, Vol. 80, 1920, p. 153). Also see Prof. Eddington's book, "Space, time and gravitation" (Cambridge, 1920), Ch. 7. The observation verifies the deduction (*b*) of the next page.

$$G_{\mu\nu} = -\sum_{\alpha=1}^{4} \frac{\partial}{\partial x_\alpha} \left\{\mu\nu, \alpha\right\} + \sum_{\alpha=1}^{4} \sum_{\beta=1}^{4} \left\{\mu\alpha, \beta\right\} \left\{\nu\beta, \alpha\right\}$$

$$+ \frac{\partial^2}{\partial x_\mu \partial x_\nu} \log \sqrt{g} - \sum_{\alpha=1}^{4} \left\{\mu\nu, \alpha\right\} \frac{\partial}{\partial x_\alpha} \log \sqrt{g},$$

$$\left\{\mu\nu, \lambda\right\} = \tfrac{1}{2} \sum_{\alpha=1}^{4} g^{\lambda\alpha} \left(\frac{\partial g_{\mu\alpha}}{\partial x_\nu} + \frac{\partial g_{\nu\alpha}}{\partial x_\mu} - \frac{\partial g_{\mu\nu}}{\partial x_\alpha} \right),$$

$-gg^{\mu\nu}$ = the minor of $g_{\mu\nu}$ in the determinant $-g$ of four rows and columns formed by the elements $g_{\mu\nu}$, and $K_{\mu\nu}$ depends only on physical quantities, *viz.* the distribution of matter and the state of motion in the region under consideration. The equations of motion of a particle, subject to inertia *and* gravitation, are given by

$$\delta \int ds = 0$$

and they are

$$\frac{d^2 x_\alpha}{ds^2} = -\sum_{\mu=1}^{4} \sum_{\nu=1}^{4} \left\{\mu\nu, \alpha\right\} \frac{dx_\mu}{ds} \frac{dx_\nu}{ds}, \qquad \text{(Y)}$$

α having the values 1, 2, 3, 4. As a first *approximation*, (X) and (Y) give the results of Newton's theory, *viz.* Laplace's or Poisson's equation, and the equations of motion of Newtonian mechanics.

There are three deductions from Einstein's theory which have attracted great attention:

(*a*) As a second approximation, (X) and (Y) lead to a small variation from the Newton—Kepler motion of a planet in its orbit, so that the angle described by the radius sun-planet between one perihelion and the next exceeds 2π by

$$\frac{24\pi^3 a^2}{T^2 c^2 (1 - e^2)}.$$

Thus for Mercury the perihelial movement is given by the above formula to be at the rate of 43″ per century, but for the other planets of the solar system the movement is too small to be detected.

(*b*) A ray of light which passes the sun at a distance of N sun-radii from its centre, should be deflected towards the sun's centre by

$$\frac{1 \cdot 7''}{N}.$$

(*c*) An atom absorbs or emits light of a frequency which is dependent on the potential of the gravitational field in which it is situated, so that the spectral lines in the light coming to us from the sun should appear shifted towards the red end of the spectrum.

The first deduction of Einstein completely accounts for the observed perihelial movement of Mercury; the second was verified on the occasion of the total solar eclipse of the 29th May, 1919; and, as regards the third, at present the consensus of opinion is, that it is impossible to assert definitely whether the predicted shift has been confirmed by observation or not.

Among the researches that followed, or were almost contemporaneous with, the publication of Einstein's memoir a prominent place is occupied by those of Hilbert[73] who gave an alternative method for deriving the fundamental equations of the general theory of relativity by using the postulate that all the laws of physical phenomena can be summed up in a generalized principle of stationary action, *viz.*

$$\delta \int H \sqrt{g}\, d\tau = 0$$

where H is an invariant. Hilbert's method has been developed by Lorentz[74] and Weyl[75] who has succeeded in welding together electricity and gravitation into one geometry. An interesting discussion of the methods of Einstein and Hilbert as well as of Einstein's conception[76] of a finite space of spherical or cylindrical shape will be found in a series of papers[77] by Klein who has shown the connection between the ideas underlying the investigations on relativity and the ideas given long before in his Erlangen programme. A well-reasoned criticism of Einstein's theory was given by Dr. L. Silberstein in a paper[78] published in July, 1918. Among the latest papers on relativity may be mentioned those of Professors A. Palatini[79] and E. Kasner[80].

---

[73] "Die Grundlagen der Physik I" (*Gött. Nachr.*, 1915), communicated on Nov. 20, 1915; "Die Grundlagen der Physik II" (*Gött. Nachr.*, 1917).

[74] "On Einstein's theory of Gravitation" (*Proceedings R. A. Amst.*, Vol. 19, 1917).

[75] See his papers in *Berlin. Sitzungsberichte*, 30th May, 1918, and in *Ann. d. Phys.*, Vol. 54, 1917, and Vol. 59, 1919. In this connection also see his book "Raum-Zeit-Materie" (Julius Springer, Berlin, 3rd edition, 1920).

[76] "Kosmologische Betrachtungen zur allgemeinen Relativitätstheorie" (*Berlin. Sitzungsberichte*, 17th Feb., 1917).

[77] "Zu Hilberts erster Note über die Grundlagen der Physik" (*Gött. Nachr.*, 1918); "Über die Differentialgesetze für die Erhaltung von Impuls und Energie in der Einsteinschen Gravitationstheorie" (*Gött. Nachr.*, 1918); "Über die Integralform der Erhaltungssätze und die Theorie der räumlich geschlossenen Welt" (*Gött. Nachr.*, 1918).

[78] "General relativity without the equivalence hypothesis" (*Phil. Mag.*, Vol. 36).

[79] "Deduzione invariantiva delle equazioni gravitazionali dal principio di Hamilton" (*Rendiconti C. M. P.*, t. 43, 1919).

[80] "Einstein's Theory of Gravitation" (*A. J. M.*, Vol. 43, No. 1, 1921).

C. G. Röder G. m. b. H., Leipzig. 821223